ENVIRONMENTAL PRESENCE©

By C.C. Krepick

Spiritual Environmentalism, where science meets spiritual Earth…

Based on the script for the TV Series/Documentary

Environmental Presence©/Environmental Presence International©

Scientists can tell us a lot about what toxic contamination is in our environment and how to clean toxins from land, air, and water. Different indigenous cultures use various methods in their historical, cultural beliefs with cleansing or healing the earth using a spiritual connection to Earth. Let's go where science meets spiritual Earth. Interviewing Environmental Scientists and people that are specialized in indigenous historical various cultures, learning about natural and unnatural contaminations in our environment. Learning how these contaminants affect the environment, people, and animals will open the eyes of the audience about how important it is to keep our planet as clean as possible from toxins. When the premise of this documentary/TV series began it also included a look at the paranormal. The choice to delete the paranormal from this idea came about as the line between spirituality and the paranormal is such a fine one. Spirituality is as real to the people that practice on their own as the Earth is to the scientists that study it, both are connected, spirituality and the Earth. Everything is connected.

Hosts will take a unique adventure investigating the mysteries surrounding environmental damage and cleanup efforts across the United States in the TV series Environmental Presence. Beginning by interviewing eyewitnesses, then specialists, and spiritual leaders as well. We all know that there is science to it all when it comes to toxic damage and cleanup efforts, but indigenous people and people of various religions have performed their own efforts for cleaning Earth as well.   This is a travel excursion into the world of toxic contamination, man-made and natural. From the city of Centralia, Pennsylvania where the underground fire is still burning to the Le Brea Tar Pits in Los Angeles, California, with some man-made toxic damage in between. There is a lot to see and discover interviewing environmental scientists, specialists, Native American Indians, Voodoo Leaders and more on how the Earth may be restored. Spiritual Environmentalism, where science meets spiritual Earth… Environmental Presence

A review of the synopsis from the late Betty Dravis. She was an incredible journalist and touched the lives of many. "This synopsis made me think a lot of things... Green... saving environment and SPIRITS ... I love this synopsis and it makes me stop and wonder...even marvel at how you have worked spirits (paranormal) into it. I would view it from this wording... More later. Wow! So very impressed with you and all you do... But very sorry for your illnesses... Dang!

Can't have it all, can we? ♥ "

Foreword

Spencer Gray

Theoretical Physics Major

Minor in Philosophy

If the paranormal was so simple we could understand it, then the human mind would be so simple that we could not understand it. The paranormal is big business right now. We have TV shows, podcasts, "ghost hunting equipment" for sale and countless books on the subject. Even the scientific community will mock the paranormal yet make money off its mystery to further research that's floundering. At the end of the day, paranormal investigations and mystics don't fork over much evidence. Well, there's the occasionally creepy EVP and maybe an orb in a photo. I hate to burst your bubble, but orbs are nothing more than a result of space-time bending during the materialization of information during a paranormal event and electrons emitting photons. Ergo this is just one of many painful truths that won't sell to or excite the curious hearts of mankind.

Don't stone me to death yet, am I saying the paranormal or supernatural aren't real? Quite the contrary! They are real, however, there are sciences like physics, mathematics and biocentrism behind them. Of course, in my opinion, a God who is one hell of an engineer and physicist.

It makes me chuckle that we are simply ants in the primordial stew but have the hubris to discount creation. For example, how do we know the theory of infinite multiverses isn't what we were taught in Sunday school to call "Heaven'? In John 14:2 Jesus promises, "In my Father's house are many mansions: if it were not so, I would have told you. I go to prepare a place for you." Am I trying to be a preacher or shove my religion down your throat? God no! My intention is to get you thinking in new ways.

After all, Socrates did tell us true wisdom is in knowing we know nothing. There is also the painful truth that no matter what, there will always be things we are not meant to know. Alan Watts said, "The further we look out with our telescopes and the deeper we look with our microscopes, the larger and larger and smaller and smaller the universe or Earth have to become to escape our investigation because we're the universe looking at itself."

And as C.G. Jung believed that above and beneath our subconscious mind flows a vast, deep sea of dream images and forgotten lore that he called "the collective unconscious", this psychic ocean doesn't belong to any one person but it was a dimension shared by all of us where we discover the amazing archetypes of mythology, demons, angels and gods. Not only does this help confirm my statement about the multiverses but it relates directly to our planet as well. Our planet exists not only because of "collective unconscious" but *because* we are conscious.

And we are pissing in this psychic ocean and our own gene pool by poisoning our world with fracking and many other corporate hazards. We carelessly rape and plunder resources with the idea that someday we'll have the technology to relocate to Mars or make a success of space mining. I have news for the ubiquitous corporate gangsters and pirates of the world. This universe is going to freeze in as little as another thousand years. It can't be escaped. The only way to avoid the end of everything is to leave the universe. However, first we need to heal the Earth. Mother earth does have a soul. She is alive just like we are. She can help us figure out our salvation but not if she's dying. She can be healed and cleansed spiritually and physically just like our own bodies. Take the hand of my friend and colleague C.C. Krepick and allow her to walk you through a spirit world of new thoughts and revolutionary healing.

*All opinions expressed in the Foreword are the intellectual property of Spencer Gray.

My interest in writing this book is so that people could better understand why the TV series that I had written was so very important to me.

I survived a brain hemorrhage from a brain aneurysm that nearly took my life in the year 2000. The completed results were that along with this brain aneurysm there were four more that had to be repaired after the one had hemorrhaged. Today, I still have one in my brain that could not be reached all those years ago, high blood pressure, daily chronic headaches, and memory issues from the hemorrhage. Back then though, I wanted to learn more on reasons that this happened to me. Now, brain aneurysms are said

13

to be hereditary, however not one single person in my family history had been subject to even one that we were aware. So, where in the world did the aneurysms in my brain come from.?.

I began to research shortly after the second brain surgery. I was a smoker, could it have been those chemicals…possibly. I grew up in a town that had been contaminated with Dioxin. Could it have been that…possibly? What other health problems had I had over the years…breast cysts at 16 years old, 5 brain aneurysms at 29 years old, endometriosis resulting with hysterectomy in my 30's…hmmm. Throughout those years, I watched my dad go through prostate cancer and is still living with near crippling at times rheumatoid arthritis, my mom had severe fibromyalgia, we just lost her from what started as a spot of Cancer in her colon to it spreading to her liver and further throughout her body. She passed November 27, 2017. All my school years, from about the 6th grade, I saw the families of my friends suffering illness and death, some at young ages. The question was always why. I know that people get illnesses that can take their life at old age. That is just the way it's been, but what is the seed that causes the illness. The body is a fantastic being, as one part to the body leads to another part of the body, so if one part fails then the next is struck with failure and so on until the body gives up.

After the last brain surgery, I had a curiosity as to what was going on with our health and I could see it went way beyond the years that I had smoked cigarettes. Some of my research is included here in small quotes between the chapters or episodes as it is. I decided in 2009 to write a documentary type script for a film or television. After I had it written and in the Writers Guild, I was having trouble getting producers to see my vision so I decided to write this book in the same layout (well, sort of) as the TV series/Documentary had been planned.

I have written this script/book with the intention to educate others on the environment around them as well as to educate on the spiritual connection that everyone has to Earth. I could think of no better way than to bring in interviews with scientists and spiritual leaders of various religions and indigenous cultures, getting the ideas from each side on what is in our environment in regard to contamination and how to clean it in scientific view and also how to heal the Earth from the spiritual standpoint as well.

It is not the intention to call out big corporations for damages or to go after them for any compensation. Choosing natural contaminations that form already in the Earth and the unnatural contaminations that have already been publicly documented by the federal government we avoid pointing fingers. Just let people know, educate them on what is happening to the land, air, and water in areas where there is something changing the area. Shed some light on contaminated areas that may be right down your street and you may not even know about them yet. Really the main goal here is to discover how these areas are "cleaned" of the toxins that damaged them. What are the steps that are taken to remove these toxins? What happens to them when they are removed? Where do they go? Then the next questions, relating to spiritual "healing" of the Earth from the toxins. How is this done? What happens to the toxins when, let's say, an Indian ceremony is performed to "heal" Earth? Knowledge is power, get everyone on the same page and make a difference. Let the people decide what we can do scientifically and/or spiritually to clean/heal our Earth, maybe even both.

I did find out over 16 years of research what contaminants were in my town and how they got there. Was I able to do anything about it, you ask. I wasn't able to get it cleaned up, though spiritually, myself and others sure prayed over it, a lot. Consider the chapters as partial episodes of a TV series or the whole book as a Documentary, your choice, but it has been my life for many, many years.

In 2012, the film The Promised Land hit theatres. Matt Damon was the actor for a character named Steve. Steve had a lot to say about the environment in this film. I remember nearly falling out of my chair when I heard him say in the film...STEVE: "Hey, did research say anything about an environmental presence?"

Script found in
https://www.scribd.com/document/113279754/Promised-Land-Script#fullscreen&from_embed

Poster photo found in
http://www.imdb.com/title/tt2091473/mediaviewer/rm4211517696

Hearing Matt Damon say the actual name of my documentary/TV series out loud like that somehow gave me a feeling of confirmation, that this is real. I am sure though that during the script writing process for Promised Land they had no idea that there was a script out there waiting for a producer to come along and carry it on to something bigger titled *Environmental Presence*. The movie with Matt Damon was based on the problem that is fracking. Environmental Presence is something else but has the Earth at its heart.

In between each chapter/episode in this book you will read anonymous quotes from people that I have encountered over the years that became ill from contamination and now are deceased. I promised them that I would let their voices be heard. This is one way that I felt that I could do it and not be harassed by the corporation that took them from us.

Christy

"Yeah, it would go out the top and it would leak out all them holes and went out over here and went out the back of the boiler because there was all kind of leaks and stuff out through there. Blowed over at the saw mill in the air, over at the kilns, and all them people over there was breathing it." ~Anonymous

Environmental Presence

Whispers of Truth

Superfund Sites are places that have been recognized by the United States Environmental Protection Agency as basically environmental disaster sites. Places that are so contaminated that the United States recognized them publicly and saw that they were so toxic that it was a danger for humans to be in the area until the areas are cleaned of the contaminants to the legal levels of toxic contamination where humans would be safe again there. Yes, there are *legal* levels of toxic contamination that are considered by the EPA as to be not a high level of danger for humans and animals.

Tar Creek in Northeast Oklahoma is a Superfund Site. Director Matt Myers wrote the script and narrated for his environmental documentary titled *Tar Creek* back in 2009. This area was contaminated in such a way that the federal government purchased the people's homes and moved them away from the area. In this documentary, professionals and many others were interviewed about the toxic nature of this area.

When mining and milling for lead and zinc was ceased what was left behind caused millions of tons of waste in this area. Acidic mine water resulted as roof collapse of the mines occurred and this affected the aquifer below. An aquifer is the source underground for where water came from for this area. Over time the acid mine water made it to the surface contaminating the surface land and waters. Levels so high of heavy metals that humans had to be removed from the area is a really, big deal. The Boone aquifer, source of water for the area, would be contaminated.

I was able to reach Matt Myers back then when I wrote the script synopsis and he agreed to be interviewed for the episode concerning Tar Creek however years have gone by now and I have not been able to reach him to let him know that this book was in process. As in some of the other chapters/episodes here you will see that the information on the sites is there but I did not follow through with meeting locals or specialists there yet. There would also be an Indian Native from the area that would show us on screen the ceremony of healing Earth. Saving that for the screen.

"It's the fact that uh, the glue that they was putting in and a lot of stuff from plywood, anything from any other part of the plant would come to them and be poured in that wood because it would be disposed of once it's burnt."

~Anonymous

Environmental Presence

Voodoo In The Air

Script Synopsis ~A Voodoo Priestess of New Orleans would be contacted to discuss history of the contaminated area that was contaminated with the Deep-Water Horizon rig when it blew up off the ocean in Baton Rouge, Louisiana. We talk about the history of the place including the contamination that took place. We discuss if she has seen that contamination has affected the people, animals, and the land in this area. Experts will be spoken to as well as locals and/or someone that worked in the cleanup efforts about which toxins are or were there and how these toxins have affected the people as they recall it. The Voodoo Priestess would be close to the area within safe distance that has been contaminated to do a ceremony as to the way that the people of Voodoo would perform to heal Earth. We learn here not only about the contamination that took place but also the cultural ways of the people of Voodoo in

Earth healing. Local people of the area may have input as well. Speaking to someone that was directly impacted by this disaster was very enlightening. Here is a write up of this mans take on his experience with the Deep-Water Horizon disaster.

James Collier
Deep Sea Oil Drilling
Crane Operator

I worked for Trans Ocean for a while, my brother worked for them as well. Trans Oceans is one of the largest drillers out there. They actually had the Deep-Water Horizon that had exploded releasing massive amounts of oil into the ocean water. Not long after I went to work for Oceaneering. Oceaneering is one of the world's largest deep-sea diving companies. I was an ROB pilot and tech in engineering. After the spill with Deep-Water Horizon, I was working for Oceaneering where we were monitoring the well after the blast and oil spill. At that time, I was working in the marine department as a crane operator under BP. It was on a rig that was the closest to the Deep-Water Horizon. This was the BP Nakika.

When I first started working offshore, at first it was really weird. I come from a military background. This kind of reminded me of grade school. The old hands working out there would be just walking through the halls, it really was weird when you stop and realize just how dangerous that it really is working out there. You learned to appreciate and realize that these guys are really taking this seriously out here to protect our lives. We had safety meetings every morning. We conducted safety drills during the week. The reason behind that is that when you drill enough it becomes muscle memory. So, in the event that a real emergency happens it would just click and you would automatically go into the mode of what you need to be doing and where you need to be. Over all I was really impressed working for these companies. These drilling and diving, ocean companies ultimately come together as a team and do projects together. Usually they are contracted by one of the large companies like BP, Shell, Exxon, Storm Energy, and other petroleum companies that are out there. Even those petroleum companies as well have very stringent standards. I had to go through special training with BP just to meet all of their safety standards.

I had to do that just to be able to step on their rig. Everything was about safety.

When the oil spill happened with Deep Water Horizon, it was one of those moments where for lack of better words you find out that a lot of all that training and importance set to safety standards is a facade. They want you safe as long as you are making money. Everything out there boils down to the dollar. When the dollar says now is not a time to be safe, then safety is not an issue. That is what happened with the BP and TransOcean Oil spill. There was a lot of factors that went into play where safety was flat out ignored because they were falling behind on the contract. When these big companies contract you out to do a job, it is just like any construction job on land you have your different contractors that had put in bids and said my company can do it for x amount of dollars and in this time frame. That is what Trans Ocean did. They dumped a bid for fulfilling that specific role and they were behind on the timeframe. I believe in the ball park of five weeks behind. People have to understand that this is a lot of money. Each day off shore in a drilling operation

the big companies are spending right around a
million dollars or more a day for this production to
go on. When you look at five weeks behind that is
a lot of money. BP wanted this job done, it should
have already been done. Things happen,
sometimes things throw you a curve ball out there.
What basically happened is BP sat on the money,
they had to wait around, ignored safety, and as a
result an explosion happened. After the explosion
they contracted Oceaneering for monitoring the
spill. BP come up with the idea to use Corexit.

Corexit causes oil to break up in tiny droplets.
It causes it to sink so that it is no longer on the
surface of the water. It had been used before in
spills over seas. What a lot of people do not know
is that this chemical Corexit was banned and is
still banned in Europe. BP knew this, BP knew
how toxic this chemical was, very toxic to humans
but also to sea life. Dolphins, fish, shrimp, sea
turtles, birds, sea gulls, to everything that comes in
contact with it. They knew before the first drop
was ever sprayed by planes or by boat before it
was ever introduced to the Gulf how toxic it was.
They knew what the repercussions were going to

34

be. They went ahead and used it anyway. You got to understand, and I know that you know a lot about chemicals and toxins, when it comes to work places whether on land or sea, any chemical that is brought into a work environment is accompanied with a sheet called an MSDS (Material Safety Data Sheet) Sheet. This data sheet lets the people handling the chemical know the toxicity of that chemical. It lets you know how flammable it is, safety precautions for using it, whether a respirator is necessary and so on. I know without a doubt that Oceaneering had to have known what this chemical was made of and how toxic that it was. Anytime that anything comes aboard a vessel or into a workplace off of a delivery truck the very first thing that gets looked at is the MSDS Sheet. So the people know exactly how to handle and store this chemical. The MSDS Sheet gets logged. Off shore it gets logged into the log books. Everyone on the vessel knows what this chemical is all about. There is no company out there that can say that they had no idea what this chemical was going to do. They knew what it was going to do and it all boiled down to money. They knew that all of this oil was spilling and they had to make it go away as fast as they could. The only way that this would be possible would be to use this highly toxic chemical, by spraying it introducing it into the ocean, breaking this oil up in tiny droplets, causing it to sink. When they made that decision, everything that I thought about them, these companies being very professional and taking our lives as top priority changed in an instant. I realized and found out how toxic this stuff was.

I asked James was there any alternative available to this chemical.

Yes, there were alternatives, the results would not have happened as fast as others. This is something that they insisted happen really fast.

So, they could have done this with something less toxic but it would take longer to handle it?

Right. It would have been more expensive also. Every bit of this could have been avoided had they followed the protocols, safety protocols that were in place. They chose to ignore them.

While I was working out there I was doing construction on BP's Nakicka. This was approximately 12 miles from the Deep-Water Horizon. We had a BP industrial hygienist technician that boarded our vessel. Me and one other person working there asked this technician if this chemical Corexit was safe. We asked if we needed to be worried about it. He told us that it was equivalent to Dawn Dishwashing Liquid and we had nothing to worry about. Now, at that time I was still thinking that this is a very professional company that is taking our well-being as top priority. I had no idea that I was being bold face lied too. It wasn't until 4 years later I get diagnosed with Interstitial Lung Disease. We were baffled, the doctors were baffled, because this is an occupational derived disease that is commonly seen in coal miner workers and farmers. With the coal miners it is called Black Lung Disease and with the farmers it is called Farmers Lung. Coal miners will get this from breathing coal dust and farmers get this disease from breathing pesticides. I have never been a coal miner nor have I ever been a farmer. I have worked offshore a large portion of my life. How

in the world would I get Interstitial Lung Disease?
… There are various forms of this disease. I got
very, very lucky, the type that I had was able to be
cured. The majority of the types of this disease
are incurable. With mine, I took a lot of daily
antibiotics, it is gone but that is not to say that it
won't come back. The chemical Corexit that was
sprayed on the gulf was banned in Europe as a
pesticide. It is a very highly toxic pesticide. It is
so toxic that when it is introduced to crude oil it
becomes 52 more times toxic. BP, Oceaneering,
these companies new this. Depending on the
temperature it can become even more volatile. We
are out there working in 120 heat degree index
with this chemical being introduced to crude oil
and they have the audacity to tell me that this is as
harmless as Dawn Dishwashing liquid. This
chemical has attacked my Central Nervous
System, now I have the trembles and shakes. I
have a slight brain damage from it. I have
constant ringing in my ears. There is a skin rash
that comes and goes on my arms and my neck. I
am basically now allergic to the sun because our
skin has natural oils in it and this chemical came in
contact with my arms, neck, and all areas of my

body. It has stripped my natural UV protection, leaving me allergic to the sun. My teeth are dying and have sensitivity in them and nasal areas are damaged. Sense of smell and taste are nearly gone. Fatigue, anxiety, and chronic pain syndrome are very bad for me after exposure to this chemical. I have been told that I may have lupus and the list goes on and on.

How long were you exposed to that chemical?

I was exposed to it close to 40 days.

So, there are other people out there that are having the same problem that you are?

I was out at ground zero, and one of the only ones that were outside for 12 hours at a time every day. Most others were inside the vessels other than the cleanup crews and people working with the spill. Yes, I inhaled, ingested, and absorbed my share of it through working there and it affected my health drastically. I was treated as if I were expendable, with all of the money that was being lost. BP, Transocean, some of these companies when they knew things were about to go south they took up special insurance so they made more money off of that well exploding in the Deep-Water Horizon incident than they did actually pumping it. The hard lesson that I learned is that they do not practice what they preach out there. When it comes down to serious environmental issues the dollar wins out every time. That oil spill out there, the Corexit is still there it will never go away.

The oil just went to the floor of the ocean, right?

Yes, they just sunk the oil. The equivalent of sweeping dirt under a rug. Now, BP puts out these nice commercials stating that they cleaned up the Gulf. When all that really happened was it was hidden and the Gulf was turned into one toxic

cesspool. Billions of gallons of the odorless, clear, toxic Corexit and dumped it in the Gulf, it's still there. People are still dying from it. You know, the 11, God bless their souls and their families, but those guys that lost their lives on the Deep-Water Horizon, were just the beginning. There have been several more that followed. This photo was from my vessel in the distance we see the controlled burn after the Deep-Water Horizon explosion.

By Source, Fair use, https://en.wikipedia.org/w/index.php?curid=49908257

Bloody Mary Voodoo Priestess of New Orleans

Mary Millan

Voodoo and nature are one; Voodoo and the spirit world are one. There are spirits in nature. Voodoo translates to mean "Spirits" Mission: To live in harmony and to keep the spirit world and the physical world in balance is a goal of this ancient animistic religion. How?

Feeding the spirits, dancing with the spirits, singing and playing music are some of the many ways that we serve the spirits known as voodoos - Loas or mysteries. These mysteries, or spirits, often have specific duties. There are families, or nations, of spirits many of them stemming from the ancient tribes which were brought to the New World with the slave trade. As different tribes were brought to different places their spirits travelled with them and each of these areas had different needs which arose creating unique interactions with the Voodoos and the individual spirits of place that they encountered.

The religion of Voodoo was, of course, also brought with the slave trade, their rituals and dance followed step. Traditions were shared. Many spirits adapted, developed and evolved. Other spirits were forgotten and faded away. The needs of an area dictated the servitude of these mysteries. Many blended, some split, others procreated and large amounts of new family members resulted. Spirits of Voodoo adopted and were adopted into the pantheon of the spirits of place. Local spirits rose up when it was time to band together and fight: to heal the Earth, to navigate the waters, to clean the air and to awaken the passions of fire needed to heat and temper the strength required. God was always in the center.

As a priestess, I have a personal relationship with the spirits. This relationship reveals what needs to be done if you are willing to listen. You can call on specific African loas who are assigned to specific areas, and use traditionally handed down methods of evoking spirits through song and gifts that are listed as their favorites. Though there is a traditional route taught to the priest and priestesses there is still direct communication with the spirits that is utterly important, Voodoo is a living religion.

Many Voodooists serve and rely on the ancestors and/or the patron spirits of their own head (met tet) or house (temple) for everything. One still needs to always listen deeper and watch one's dreams, for a form of spirit possession occurs where the spirits can instruct you on what is needed to heal the particular situation. Personally, I even get recipes in my dream visions. I may see in vision a food that needs to be served, herbs and gifts and sticks and stones. Many aspects are revealed depending on the needs.

"I wait for those special keys to unlock the right remedy for the situation. I await a vision or a dream."

I have utilized all these above methods and follow through.

In the New Orleans area, it seems that I work mostly with the water spirits. And, of course, I work through swamp magic - a sort of amphibious between-the-worlds ancient working. We are a city of waters, many rivers and bayous that are being polluted and destroyed by man with oil industry and other industries.

"I've heard the ancients cry out in pain and seen fish-people gasping."

Humans are sometimes so busy in life that they miss the big picture, and there's a big picture going on right now. The divine earths and waters need protection and government is stepping back instead of forward.

"This is the time for the voodoo and magical communities in general to step out if the shadows and perform their rituals to heal. "

Problems like this can often feel so big that the individual often feels powerless. This is the best time to turn to the spirits and the ancestors. Dance with them. Those ancestors, who I call the architects, those who have lived in a place for a long time prior, those who actually built it, have a natural interest in protecting their homes. Call the ancestors and plead their forgivingness for the arrogance of man and ask for their power to rise and work with the nature spirits to heal.

Anecdote:

After Katrina, for at least three or four years, I did healing rituals non-stop. I had to balance what was going on, on a spiritual level, to help the physical plane heal. This was a difficult process as both sides of the veil were affected heavily. Then in the midst of it all there was the gulf oil spill, followed by flooding all over the country. I traveled in the flesh and in astral to help other areas heal; from New Orleans, all the way up to Canada where I was working with other people with their oil and flooding issues. These acts and Rituals may be understood as a butterfly effect, but that butterfly effect can cause mighty hurricanes as well as mighty healing's too. I believe that mother nature and mother water, or as we call her Maman You, was crying out and is still angry now at our destruction.

"We need to beg forgiveness and show why we are still worthy of life. We have to realize that our prayers, our efforts, even a single song are important messages to the earth and water spirits."

There are those Citizens who need to speak on the physical plane and lobby the government for change. And There have always been those that are behind the scenes singing, dancing, praying pleading with the world to forgive us. Both sides matter and together they can hopefully to balance the Earth.

In Voodoo, there are a belief in deities that we honor to help the world, there are the exalted ancestors who we need to respect and cleanse. There are nature spirits we need to align ourselves with and then there are all of us: the family of humans who need to try to love each other as a family.

All need to be remembered.

All need to be fed.

All need healing.

"We did things that we did not know whether they was right or wrong back then. We didn't know it wrong, we did what they told us."
~Anonymous

Environmental Presence

A New Day

Script Synopsis ~ Host contacts a local psychic of Camilla to discuss history of the previously contaminated area of Camilla Wood Preserving Company in Camilla, Georgia, which is now a soccer field for children. We talk about the history of the place including the contamination that took place. We discuss if she/he has seen that contamination has affected the people, animals, and the land in this area. What would a psychic do to heal the Earth? Is there a prayer or a type of ceremony that she would do? Host speaks to experts, locals and/or someone that worked the case about which toxins are there and how these toxins have affected the people as they recall it.

The following was gathered from the Deccan Herald online. This online article discusses different toxins in relation to air pollution and handling waste. Ankita Sah is my friend from

51

India of whom I admire deeply for her loving spirit and interest in the beauty of nature. Ankita granted me permission to use her photo of the Pine forests, Ranikhet, Uttarakhand — in Nainital,

India. This is the forest that you see on the book cover. It is so beautiful.

Bio-medical waste: What lies in store?

Written by researchers at the Council on Energy, Environment and Water, New Delhi,

India, Ankita Sah and Vaibhav Chaturvedi, April 03, 2016

The government hopes that the new rule would change the way of waste management in the country and thereby also prove instrumental in the 'Swachh Bharat' mission. Bio-medical waste comprises of human and animal anatomical waste, along with treatment apparatus like needles and syringes, as well as discarded medicines and cytotoxic drugs among other things that are discarded as waste from hospitals, nursing homes, pathological laboratories, blood bank, among others. Management of this waste is a challenge for policymakers and healthcare managers alike. The new rules in general are a definitive improvement for better management of bio-medical waste. However, one specific aspect needs a special mention, and that is a clear effort by the government to reduce the exposure and negative health impacts due to emissions from bio-medical waste treatment process. This is evident from two aspects of the new draft rules, a limit on the emissions of dioxins and furans, as well as more stringent emission standard for particulate matter; and a strong regulation on on-site opening of any new common bio-medical waste treatment and disposal facility (CBWTF). Dioxins and furans are toxic/carcinogenic emissions and can seriously impact the health of humans and wildlife that are exposed to these emissions from the waste treatment facilities. These are generally not man-made substances and are listed as persistent organic pollutants (PoPs) in the Stockholm Convention ratified by India in 2006. It is therefore essential for signatory countries to take adequate measures to eliminate and reduce where not possible to eliminate, all sources of dioxin and furans. These toxic emissions are released during the incineration of medical waste. There are different ways to handle waste, but incineration is imperative for some specific kind of bio-medical waste like bodily fluids and anatomical waste. As per some estimates, emissions from hazardous waste and biomedical waste incinerators in India contribute to about 64% of the estimated dioxins/furans released into the air. The seriousness of impacts from bio-medical waste incineration can be gauged by the fact that there is an ongoing court case in the Delhi High Court on this matter. The residents of Sukhdev Vihar locality in Delhi a few years back filed a case in the Delhi High Court seeking closure of a bio-medical waste treatment plant established in their vicinity in 2006. The residents claim that the plant burns 12-15 tonnes of bio-medical waste exposing 10 lakhs residents around it to harmful emissions. On the other hand, the Delhi Pollution Control Board estimated that the total bio-medical waste generated in 2010 was 10.125 tonnes/day. There are 10 incinerators across Delhi including the ones at three CBWTFs. To put things in perspective, according to the government there are a total of 198 CBWTF operational in India and 28 are under construction. Some 21,870 Health Care Facilities (HCF) have their own treatment facilities and 1,31,837 HCFs are using the CBWTFs. As CBWTF, health care facilities and hospitals are generally located in populated areas, there is clearly a significant exposure of human beings living around the CBWTF that the new draft rules seek to reduce. The earlier bio-medical waste management rules of 2011 prescribed no limits on the emissions of dioxins and furans. The new draft rules in this sense are a landmark, as these specify a limit to the emissions of these toxic substances.

The new rules have prescribed emission standards for existing incinerators for dioxins and furans to be 0.1 nano-grams toxic equivalency/ normal cubic meter within two years from the date of commencement of these rules. Though these standards don't appear to be as stringent as in some developed countries, these are still a significant improvement over existing practice. In addition, the emission limit for particulate matter has also been lowered from 150 micro-grams / normal cubic meter (mg/nm3) to 50 mg/nm3, a significant reduction. In another important step related to the opening of new CBWTF, the rules state that no occupier shall establish on-site treatment and disposal facility, if a service of common bio-medical waste treatment facility is available at a distance of 75 km and in cases where this is viable it is imposed that the occupier must set up requisite biomedical waste treatment equipment like incinerator, autoclave or microwave, shredder prior to commencement of its operation, as per the authorization given by the prescribed authority.

Air pollution a concern

The air pollution and health impacts debate in India has been dominated by emissions from large point sources like power plants as well as by emissions from the transportation sector. Toxic emissions from incineration of bio-medical waste is another source that the Indian civil society needs to be aware of. Though in terms of the larger picture the contribution of bio-medical waste management in total local pollutant emissions could be low, the localized impacts could be high as dioxins and furans are highly toxic particles. The new draft rules are a very positive step taken by the government for addressing this concern and will ultimately benefit the millions living around bio-medical waste management facilities.

(The authors are researchers at the Council on Energy, Environment and Water, New Delhi)

- See more at:

http://m.deccanherald.com/articles.php?name=http%3A%2F%2Fwww.deccanherald.com%2Fcontent%2F538176%2Fbio-medical-waste-lies-store.html#sthash.wddnHMR3.dpuf

"So, they had me over there, I lasted a year or two. I was over there on a dozer. I would go out there and reblade that stuff up and uh, in the heat, you could just feel it burn your throat."

~Anonymous

Environmental Presence

Whichever Way The Wind Blows

Script Synopsis ~ Host contacts a member of local (Klamath) Indian Tribe to discuss history of the previously contaminated area of The Taylor Lumber and Treating Superfund site located in Sheridan, Oregon. At the site, Taylor Lumber and Treating (TLT) conducted wood-treating operations from 1966 to 2001. The rest of the episode follows the premise of previous episodes.

From what I have learned through research most all lumber mills at some point have done about all of the processes for treating lumber as well at some point in time. As I see it, they have probably all contaminated the Earth the same way as any other lumber mill may have. The processes that involve getting lumber to your local lumber store where you buy it have been done at every lumber mill whether they say it has or not. Thing is, it isn't always their fault. It may be deadly what happens to our environment from the lumber process but they do not always know what they do… The lumber business, say, back in the 1900's had absolutely no idea I am sure that they were handling materials that could kill people and animals if the materials were not handled properly. As the years move on things change in the way chemicals are handled and new chemicals are created to do much of the same thing as the old chemicals were used for. Now many, many years later we know the danger, rules are set in place. Most follow them, though some do not. The reason that they wouldn't is because the "safety" measures of using such chemicals could be costly. The days of, "I'll be long gone before any damage is noticed." are over.

We travel to Sheridan, Oregon, to see what had to be done to clean the area of contaminants. Ask the right questions that will educate people regarding what toxins were there that were such a hazard to humans and animals. How were these chemicals cleaned from the area and where did they go in the process of cleaning it? What is the process of removing this type of chemical that was found here? What did the Indian tribe found in this area do to spiritually clean the Earth, air, and water from the toxins?

"Everything, anything that was hazardous came to the fuel yard because it was to be burnt and it was disposed of when it was burnt, it was goooone." ~Anonymous

Environmental Presence

The Golden Light

Script Synopsis ~ Host contacts a local Gold Minor to discuss the history of Gold Mining in Northern California regarding mercury contamination of ground and water, which results from it. We watch the old-fashioned way of mining for gold in creek/river water by example. People have died mining for gold. The rest of the episode follows the premise of previous episodes.

Would there possibly be a natural mercury problem from mining for gold? Would be an interesting adventure to actually go there, talk to people, and learn more about this.

Candy
O'Donnell
Published
Book Author
Psychic

Helping to
heal the earth
Harnessing
healing energies
from the earth is one way to ensure a thorough
cleansing. I personally like to walk the route of the
Native American as I ask the creatures and other
spiritual beings such as ancestors to assist.
Burning incense, meditating and praying for the
Earth to be renewed is one of the many ways I
heal.

I believe in a higher deity, and I ask that
elevated energies come in to allow for the negative
to dissipate. This will only occur if one has
positive intentions and unconditional love to give.
I also have psychic energy that I believe mixes
with healing energies which assist to the core of
Earth's issues. I always do this with a open heart
and an open mind. Our planet deserves only the

best and I offer up my higher self whenever I perform any healing.

"I'm gonna say he knew how to get by with it cause the EPA was coming in there and he knew how to take care of everything. Man, he knew what to tell them and he had us take care of the stuff before they got there." ~Anonymous

Environmental Presence

Something Wicked This Way Comes

Script Synopsis ~ Christy contacts a local Denver Pagan for help in the community of Arvada, Northern Colorado. This town has history of radioactive contamination to air, ground, and water in and around the Rocky Flats Plant. Cancer Clusters and many have died in the area. The rest of the episode follows the premise of previous episodes.

Bruce Jr. Barraclough-Pagan

Myself and the Pagan path has been a long one some will say a hidden one for many years. Growing up, I was always attached to our planet. I would always feel the need to be in our forests, woods, camping, hiking, just sitting there breathing the wonderful air. I always felt as if this energy would call to me, that this energy wanted me to be there.

As a child, most kids were into video games and sports, but me, I was a part of something more and not till I was around 16 did I know this for sure. When I turned 16, I received an old box felted inside filled with rune stones, tarot cards, a book filled with information, handed down crystals and stone, and a letter to me, stating I am chosen to lead the bloodline which was for many years in my family. I then started putting two and two together. I was told I had much more to learn and a path now to follow that I had a strong Keltic/Celtic background.

The Earth, to us is one in the same. To what we need to live it is a foundation that our religion was built on. It to us is very sacred for a number of reasons, as is our moon and sun. The earth is what our feet touch every day. The energy that one feels when walking on soil and the grass while barefoot, the air gently brushing against our faces, we take a deep breath and feel relaxed. The Earth gives us much in food, water, and supplies us with what we need for shelter. Without soil for trees to grow high and strong we would not have the air that we breathe. Without the wonderful soil for plants to grow we would not have food for us to eat. When all is combined if you take a minute to go in your backyard, no shoes, no socks, feet on the ground, take in the air look up at the moon or sun you will

feel as we have felt many years ago, without all
this technology to fill our hands, cluttering our
minds, you will and shall feel relaxed and at
peace. Even with what life the Earth brings to us,
it also brings life to many other species of animals
such as deer, bear, racoon, and birds us many
years ago, had much need for these creatures that
the good ole Mother Earth has offered to us, one
for food, again much protein and we used antlers
and bone for hunting and pottery some of our
everyday needs. The Earth has also supplied these
creatures with an energy to thrive and live as well
as to supply us with what we needed and in return
we supplied the Earth at one point in time life
itself. Whether that be us or some of the amazing
and beautiful creatures, one example would be
bees as they go from one flower or plant to the
next traveling many miles to spread the pollen.
Other animals, like squirrels and birds spreading
seeds from one area to another. We as humans
many years ago, understood the need of life and
Earth. Myself as a pagan, hopes one day to bring
that back.

When we say "heal", we mean to "restore", fix
something that is broken, to bring back integrity.

We all as a race together can help rid this world of the toxins whether you're a Pagan, Druid, catholic, or Christian, remember we as a race are most of the cause of why this world/Earth is where it is at now. Myself being a Pagan and some of the steps we do and take to help and even some rituals we might perform, I will explain. First and foremost be mindful of your trash and waste. Let's not just throw it on the ground when clearly there is a trash bag every 20 feet, if possible try walking a bit to work or the store instead of using your car and gas. Let's not leave the water running or dripping for hours when not needed. Let's not waste food when we don't have too… looking at all of this it becomes a downward spiral. We need oil which comes from the Earth to drive our cars, boats, and bikes. We harvest more food than needed. When we choose to order the pound of chicken, you end up knowing your only eating half. We use up much more water in a forty-minute shower when all that was needed is twenty minutes. All of this is on our planet for us to use but not to abuse. We take for granted these things. Remember, you can also go to your local forest pick up garbage. You might be able to help purify the waters that run through. Now for me and my faith and some rituals, I might add to that I will perform. Remember all magical and spiritual practice is based on the understanding of the fact that realities and energies exist outside of what we can see, smell, taste, hear, and touch, something that we once understood but since then have lost. Remember our energy is one in the same to the Earths energy. We feed off of each other. I like to go for a walk in the woods where I am closest to nature, play some fine tunes to lift the mood, find a beautiful spot to sit and meditate. Light some candles, some like the idea of colors. I usually use only white, as remember we did not have all sorts of colors for our candles years ago. It was the intent that you put into your items and I shall add some incense. I will bring with me an offering for the gods and goddesses, such as an apple, nuts, berries, and place them by a tree. I may pray for a bit, in my prayer as I call to the goddess say certain words ask for purity and cleansing of the Earth, call to the elements for help to do so some sage, white and lavender are good to use. Set the sage out in the area and share my energy. Doing this, I could receive the same. Most of all remember, just be mindful and teach, help one learn, what you can do to keep mother Earth going strong because without mother Earth there shall be no us.

"Yeah, it would go down through the screws then up through, into the fire and we…they said it would reach a temperature high enough that it wouldn't hurt nothing. We complained about it. Got hot enough that it would destroy any kind of dangerous stuff. Which sometimes we wasn't running near as hot with that wet stuff as we normally could have kept a fire with shavings. We could run a high temperature with shavings ya know. But you couldn't run it with anything wet and this glue stuff is wet. Ya know, glue water is what they call it." ~Anonymous

Environmental Presence

Natural Tar Pits

Script Synopsis ~ Host contacts La Brea Tar Pits in Los Angeles, California. Host consults with experts there to discuss history of what natural toxins arise from these pits and how it affects the area if at all. Many, many animals died in these pits preserving the animals forever. What is the difference between this type of natural tar and the tar that we use on the roofs of our homes as well on the highways? The rest of the episode follows the premise of previous episodes.

"And it goes down about a mile in the Earth?
Well, it goes a long ways. Matter of fact, when we
was up there one time, they had a fire up there.
You had to be careful cause the fire would burn a
pit up under the ground, like a cave there. That's
what scared me. That I could have just went off in
that whole where it had burned out. ~Anonymous

Environmental Presence

Don't Fall In

Script Synopsis ~ Host contacts a historian on the area of Sour Lake, Texas. This town was originally named Sour Lake Springs, after the Sulphurus spring water that flowed into the nearby lake; the Sulphur was a sign of the crude oil that lay in proximity to local groundwater. This city contains one of the largest sinkholes in Texas. What creates a sinkhole, could it be drilling for oil? The rest of the episode follows the premise of previous episodes.

"You know we used to live right there in town in, right by the creek. My kids would go over there and catch fish and stuff like that. I did not know that they was running that stuff through the creek. You know they would just get in there and play in it. After later on we found out what was going on and it was too late. We would wash clothes, we couldn't hang it outside, it would get black from all the soot coming down."

~Anonymous

Environmental Presence

The Place on The Edge

Script Synopsis ~ Host contacts a Fairy
member from the area of Tampa to join in on the
Pam Callahan Nature Preserve in Tampa, Florida.
Yes, some people believe in fairies and some
believe that they are a fairy. She consults with
experts to discuss history of the area as well as
documentation of toxins having been found in the
nature preserve. How do the toxins found affect
this nature preserve and how did they get there?
How were they removed and where did they go
upon removal? The rest of the episode follows the
premise of previous episodes.

"No, just from the machines leaking, gearboxes leaking, soaks down through the concrete somehow and comes out down there at the creek. You can just about tell where it comes out. But you gotta be standing right there close. If they won't let you in there you can't see it. But when it is not running like it is right now you will see a film all over it, the water. Now turtles, I think they go down or up or something. You get hot summertime like this you don't ever see any of them dead but them little old fish stay in there will start floating to the top." Anonymous

Environmental Presence

The Real Silent Hill

Script Synopsis ~ Host speaks with locals and psychics in areas around Centralia, PA. She consults with experts to discuss history of what natural toxins arise from these ground air vents releasing steam and smoke into the atmosphere coming from the burning underground in the mines. What caught the underground caves on fire and why does it still burn? How will this tragedy affect the Earth in years to come? The rest of the episode follows the premise of previous episodes.

I wanted to add an Indians point of view when it related to Earth Healing. I spoke to my friend Tony and he was reluctant. He said that what they do is private to the individual tribes. He wanted to educate on the subject so he would tell me what he could. Below is his view.

Tony Wolf Paw-Apache

The Earth and ourselves have a union that is interpreted in many different ways. It all comes down to the same perspective, just said differently. The wounds of Earth are like our own. The Earth has the markings of all its ages. Just like we have all the ages that we have lived. The Earth has its light and dark side like we all know... just like us. We live on the surface of a living system, that is part of a living system. Living systems communicate with each other. We acknowledge the living system, and the living system acknowledges us... like looking into a mirror... you move, I move. We as humans in daily life have lost the perspective of living. We lost sight of purpose and we have lost connection with the living system. We are and always have been part of this living system. Through ceremony we acknowledge and reaffirm our connection.

My Indian given name Wolf Paw was given to me by my Elder and Chief of Clan Nilchinaa, we called him WindTalker. He crossed over few years ago. We are Apache. His ceremonies reflected his unity and communion with Earth and Sky. Mine does as well. When people ask me to explain my ceremony, it's like they are asking to explain why I love my wife and kids. It comes from within. Ceremonies are very simple. Many acts in a solemn way are symbolic... both traditional and personal ways. You can write a whole book based on ceremonies alone.

Some tribes leave their homes all gathering away from sight of city lights to worship and do their ceremonies. Some I have not seen myself. When you're invited to one, it is an honor.

I believe in good medicine both for spirit and body. You have to be present, for most are not shared like this. It comes from elders who bestowed their knowledge to others. Others, I can share because I believe it doesn't belong to me but for all. The reason why we are unbalanced is because we become selfish in our way. Where I stand, I take my medicine with me. I find myself in constant prayer and chant. Not because I'm a medicine man, like others think, only because I need it, I too have my doubts and struggles. I keep my rhythm with the rhythm of everything around me. I believe that if I heal myself, my surroundings will see the affect... and if the Earth heals... then it will reflect on me too... it goes both ways. That's why we call it sacred. We can't separate it. Selfish and unbalanced… If I go by not caring about things... it reflects in my surroundings, the environment talks back. We sing to it. Drum and dance reflects our unity with the rhythm of everything around us. We are born again. Everything talks back. We need to learn simply how to listen and to accept what we are being told. We must listen to the words of our surroundings. Good observation, the words will come naturally to you. You will know it yourself. You talk with your environment... You are talking with creator, creator of all creations.

Not just the one you choose, because that will be a selfish conversation. You surrender your personal needs and selflessness to accept what you need... not what you want. Ceremonies are made to help guide you to a personal journey. Your life... becomes that ceremony. Your given name becomes you, your colors become your life, your dance becomes your rhythm, all is one.

Some people say... I would like to know how to dance... and they copy the dance of others... not really finding their own rhythm. If I do what others do... more likely I will make the same mistakes and pass those on as well. Life is about discovering who you are.

Our nation has a medicine man, and I personally know two more. I am both Arawak and Ndeh working as secretary of state with the goal of establishing an indigenous citizenship away from the restrictions of reservations and other policy.

Native ceremonies never came with an instruction book. It was by observation and acceptance. I have argued about ceremony with other tribal medicine men and leaders in the past... my views have not changed. They remain the same. Others see it different because they keep things guarded in a selfish way. That is not a ceremony anymore but a show. You can't heal selflessness... that is a disease. That is what ceremony is called, to cure.

As you can see here, I did not include every episode that I have available. There are 21 in total. There is also a spinoff television series titled Environmental Presence International. There is so much to learn about Environmental Toxins. The different toxins that are considered to be the worst for our environment are the toxins that we all as humans need to understand the most. Where are they coming from? How can we clean the Earth scientifically? Is there something spiritual that can be done to heal the Earth? It takes experts in environmental science and spiritual leaders to answer these questions. The common bond that we all have is that we need the Earth, water, and air, to survive. If these things are going to be damaged due to progression then we need to understand what it takes to clean it up. It should not be that only specialists know this information. Every person should have the opportunity to learn what it takes preserve our environment for healthy lives.

Something else, there is but one Creator. To me myself, that is God. To others he has a different name but nonetheless it is still the one Creator. Learning about the way that other people believe that they can heal the Earth is no more than their own way of prayer, be it a ceremony or whatever. We all have to learn and stand together no matter the religion or cultural belief.

I do hope that in the future you will see this television series come to life. People are beginning to wake up and see the changes that our Earth is making. The changes that are happening are sometimes horrific for us humans and animals. Maybe it is the Earth's way of saying that it is bigger than us and we are not respecting it anymore. The old saying of take care of what God gives you comes to my mind. The future is in education about the Earth, water, and air. Without those three powerful things we cannot live.

Write from your heart, what you believe in your heart, the people will hear your heartbeat when they read your words. Your words will breathe life into their bodies, empower them, educate them, and make them see the world in a different way. ~C.C. Krepick

A foundation that I support however I can is the Ian Somerhalder Foundation. Everyone that I speak to about these topics I mention this foundation to them. To learn more about the environment and ways to contribute to a better one, please visit the IS Foundation. Get involved... "The Ian Somerhalder Foundation works to empower, educate and collaborate with people and projects to positively impact the planet and its creatures. ISF delivers unique programs and services and provides public outreach, education, and grants in support of Creatures, Environment, Youth, and Grassroots initiatives." "The Ian Somerhalder Foundation was established in November 2010."
http://www.isfoundation.com/

In Memoriam Of:

Alan Eudy

Terry Eudy

Ray Wilson

Judy Wilson

Mitchel Kesterson

Arnold McBride

Mela Rider

JR Rider

Sharon Pennington

Joan Manley

and thousands more that will never be forgotten.

C.C. Krepick, M.Ed, B.A. Eng Lit, is an American Book Author. Books published in revised edition, In The Dark With Good And Evil, and a children's book Whimsical Fairies, found in various places online, available to local bookstores through Ingram Publishing. January 2011, Christy wrote/created, produced, directed, and published her online TV show, The Bradshaw Chronicles, broadcasting on an Internet television channel to millions of people worldwide for over two years that was based in entertainment. She interviewed talent from the late actress Lisa Robin Kelly of That 70's Show to the band TESLA. Some of those shows are still archived on the Internet in various places. She had a character part in Book Author

Brian Wilson's book Return To The Moon the sequel to Brian's previous book Honeymoon On The Moon. For a short period of time Christy worked with Candy O'Donnell in an effort to help find the missing using journalism skills on CNN iReport. Christy wrote a television series in 2012 titled Environmental Presence, which is registered with the Writers Guild of America. Christy spent over 10 years studying environmental disasters and spending time with the people whom had been physically affected by them. Her most recent published book may be found in Amazon, BAM!, and Barnes & Nobles titled, It's My Life…It's Now Or Never…

http://www.cckrepick.blogspot.com

http://www.facebook.com/EnvironmentalPrese
nceTheTVSeries

http://www.facebook.com/cckrepick

www.ingramcontent.com/pod-product-compliance
Lightning Source LLC
Chambersburg PA
CBHW070155230526
45471CB00002B/668